FINALLY A REASONABLE THEORY OF EVERYTHING

PREFACE

It was the year 1977, I was taking a physics class at the *Maravillas* School in Madrid, when my teacher "Félix Marcos" (to whom I dedicate this book for getting me "the physics bug") made this comment to the whole class:

"Nobody knows why Gravity occurs"; he took a pen and dropped it from the height of his shoulders and it fell on the desk and he repeated: "Nobody knows why the pen falls, and if someone discovered it

they would give him the Nobel Prize immediately."

My eyes widened, and I wondered how we could be unaware of everything about a force that is affecting us permanently from the moment we wake up until we go to bed.

I was so impressed by it, that from that moment on, I decided to dedicate myself to investigating Gravity, not in a professional way, because this does not make you earn money, but in an amateur way.

Without the desire to discover how it works, but with the passion of curiosity; a passion that encouraged me to continue throughout my life in the gap, looking for "a reason", although it took many years and many disappointments, many Theories thrown away, until in the maturity of my life I found a Theory I am comfortable with, and I can finally rest.

I can, therefore, affirm without exaggeration that it took me 40 years to prepare this book. Each of the Theories presented here are not the result of spontaneity, but have been made on the basis of previous theories developed by myself and, that in most cases, I have discarded, because they do not fit all the standards that the experience of current physics experience gives us.

Behind each theory raised here, there are from 10 to 12 previously discarded theories.

As for whether I have the right to criticize the entire previous scientific basis, the answer is a categorical "YES", because science at present is not capable of explaining in a reasoned way the fundamental causes of the existing Forces of Nature such as Gravity, Electricity and Nuclear Force.

THE 4 FORCES OF NATURE

In current physics we describe the existence of the 4 Forces of Nature, their adequate explanation would make up the structure of the Theory of Everything.

STRONG NUCLEAR

WEAK NUCLEAR

ELECTROMAGNETISM

GRAVITY

Research has consisted of many scientists such as Einstein, hawking; in the Search for a single formula that encompasses the 4 Forces.

Personally, I understand that the last and basic explanation of the Forces must come first before the formulas. Sometimes we are capable of giving a formula for a Force, and with that it seems that scientists imply that we have found the solution and the ultimate cause when the ultimate and fundamental cause of such a Force is totally and absolutely unknown; such is the case of Einstein's Gravity behavior formula.

Although we know how gravity behaves, we are completely unaware of its ultimate and fundamental cause.

THE WEAK NUCLEAR FORCE

We can consider the Weak Nuclear Force as the radiation emitted by the change of chemical composition of the elements.

We know the following types of rays produced by radiation:

ALPHA RAYS

BETA RAYS

GAMMA RAYS

This Force is the only one that I personally understand sufficiently explained: Radiation rays produced by the chemical decomposition of elements are capable of producing alterations and especially energy capable of heating or passing through objects, etc.

But what about the other Forces.

The starting point was made up of my reading of the "String Theory", and although I have my discrepancies with it, this theory taught me to think about Natural Forces in a quantum way, this is imagining the world of very small particles that we do not see either with the naked eye nor even with the microscope.

If the apple of a tree is attracted to the ground, it is because between the apple and the ground there are a large number of tiny particles that we cannot see and that perform an attraction function between the apple and the ground.

TORNADOES

To begin with, let us look for a natural phenomenon that produces, attraction.

If you can't think of any, after "years" going around it, I remembered a natural phenomenon that produces attraction: "Tornados."

Could it be that the particle that causes gravity and that we have defined as Graviton, is actually a tornado capable of attracting what it has around it? In such a way that if 2 Gravitons intersect, they would unite their tornadoes into a more powerful one.

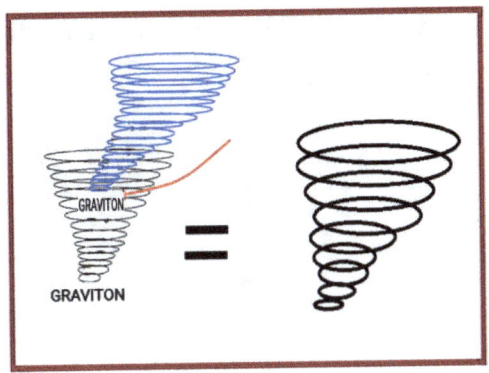

And so on indefinitely, in such a way that if a large number of Gravitons were to join we would have a Black Hole.

This system would lead us to a new problem, and it is that "Lumps" would form:

Gravitational foci would form everywhere, and this is not how we observe reality that exists in our environment.

Which brings me back to where I started.

We are going to abandon the idea of Graviton as a tornado, however we use the tornado to

see if it can be applied to another of the existing
Forces of Nature: "Electricity."

ELECTRICITY

ELECTRIC ATTRACTION:

To do this, let us imagine that proton is a rotating particle (with spin) and it creates a tornado or spiral around it, capable of attracting particles that it has in its environment.

The reasoning would be similar to that presented previously with the Graviton:

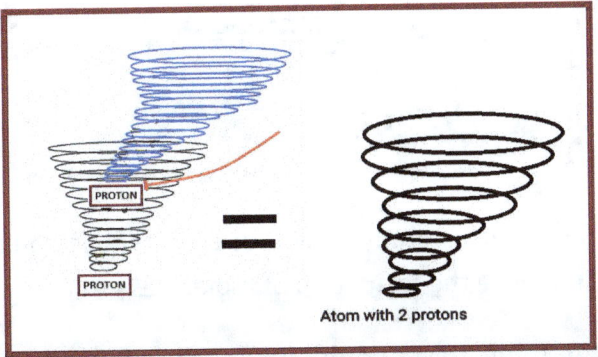

Atom with 2 protons

This would be a graph of the formation process in the stars of the element Helium by fusing 2 hydrogen atoms.

The atoms would capture in their tornado a number of electrons proportional to the force of the Tornado.

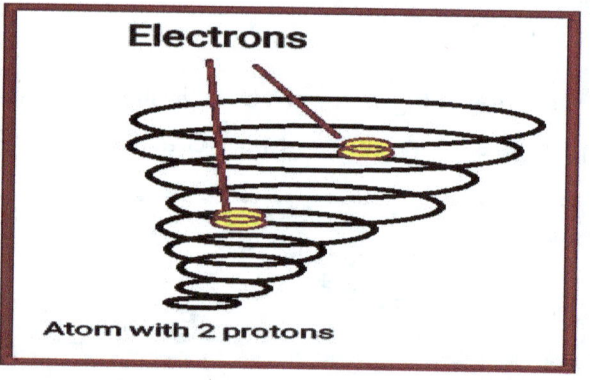

Now let us study how these Tornadoes could influence the formation of molecules:

And for this, let us look at a new atomic scheme: one in which the tornado appears more like a galactic spiral:

This would be the scheme of an atom with the protons in red and electrons in yellow.

Here the scheme of formation of a molecule by the union of 2 atoms

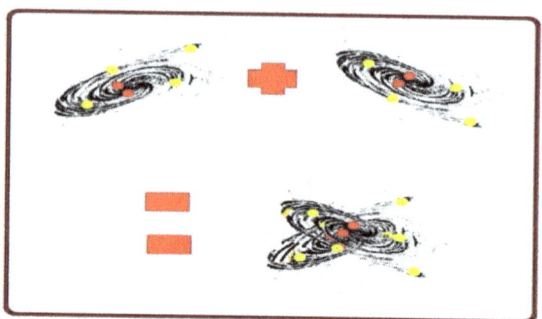

But in reality, what would hold the molecules together, since 2 tornadoes by themselves would not have to join; the essence of the electro-molecular union, would come from the perimeter of the molecule where a circular current of negative particles (electrons) would form. This circular current that would encompass the 2 atoms, would hit the outer electron layer of the atoms inside, this outer layer would rebound those hits to the inner electron layers of the 2 atoms. And the inner layer of electrons of the atoms would rebound those hits to the protons of the nucleus, which are the ones that with their rotation cause the spiral.

When the protons of the nucleus were hit by the electrons of the inner layers of the atom, they would correct their position and their spiral approaching the molecular center and keeping the molecule together.

This would be the ultimate cause of the molecular atomic electrical attraction.

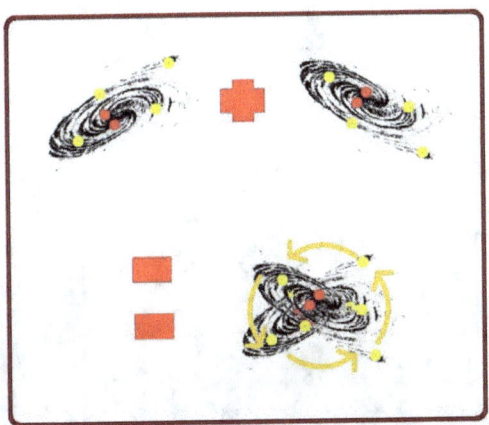

The arrows in orange represent a current of electrons that surround the molecule formed by the 2 atoms.

Let us now look at the formation scheme of a more complex molecule.

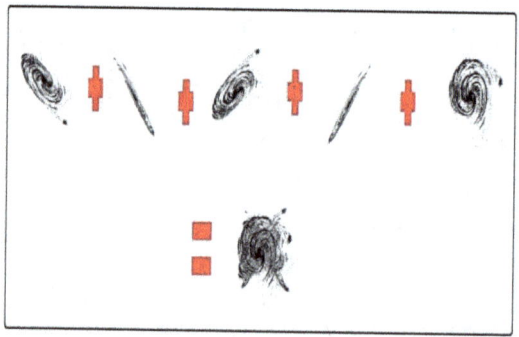

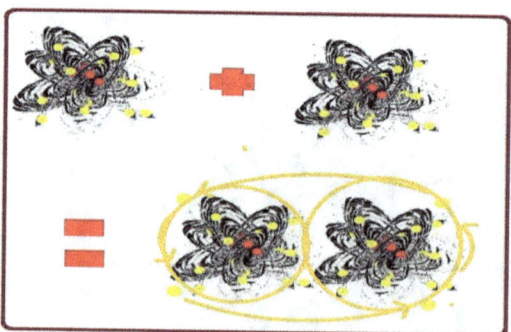

Representative scheme of the union between a molecule of a chemical element with another molecule of the same element, with the electron currents circling the exterior of the 2 joined molecules, as much as each molecule separately.

And then the union of several molecules forming a compound.

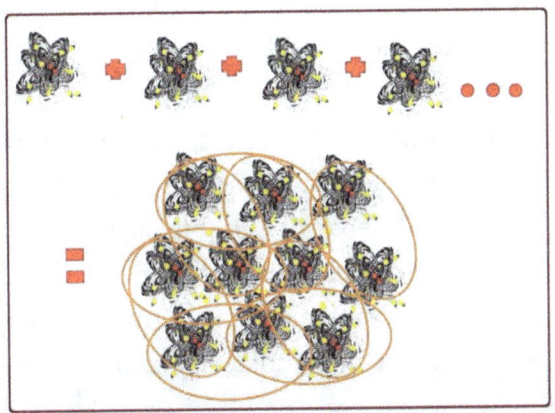

We see that not only in the periphery, but also between the molecules, currents of electrons are also formed that would maintain the cohesion of the compound in the event of partition. Thus, considering the molecular bond, we could venture a reasoning about the attraction or molecular electrical bond: In case of a compound formed by a molecular group and lacking electrons in relation to the force of its atomic tornadoes, put in contact with a compound with excess electrons, it would transmit electrons to it in such a way that electron currents would be formed between both molecular groups, and as we have explained, these currents would tend to unite both compounds, through the transmission mechanism of electron collisions in the atoms so that when they reach the nucleus, the protons also receive a sample of the collisions and move towards the molecular center.

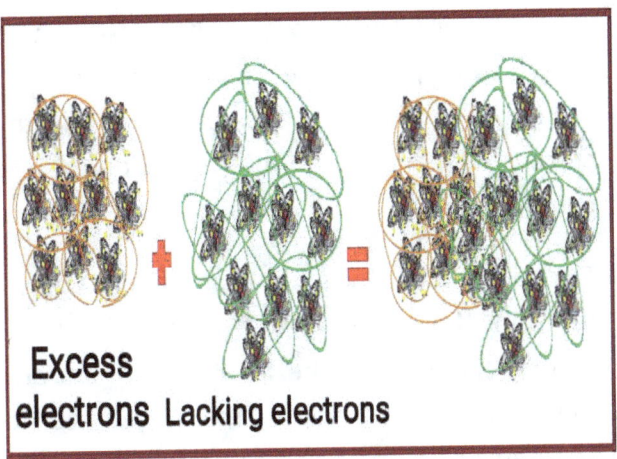

This scheme is also valid to explain an electric current passing through a conductive wire.

It seems that we are understanding how electrical attraction occurs. But, how do we now explain electrical repulsion?

If there is something more difficult to explain than attraction is repulsion and for this, I admit that I did not get an answer until years later when investigating another phenomenon that also had me in suspense.

This phenomenon was the experiment described by Ampere on attraction and repulsion of 2 cables in parallel with electric current, depending on the direction of the current.

This experiment was the cornerstone that I used in the development of the Magnetic theory, explained in the book which I am also the author: "Unveiling the Mysteries of Magnetism"

ELECTRIC REPULSION

ELECTRIC CURRENT CABLES IN PARALLEL

To see the essence of magnetism, we have to go to the most basic and most inexplicable point of the magnetic phenomenon, which is the magnetic attraction and repulsion that occurs when electric current circulates in two electricity cables located in parallel:

Ampere carried out this experiment and verified that two parallel cables in which electric current runs in the same direction attract each other and if they run in the opposite direction they repel.

To study the phenomenon, we are going to see a scheme of what happens in the nitrogen molecules in the air (nitrogen makes up 78% of air), depending on how the current circulates through the cables. **IF THE CURRENT CIRCULATES IN THE REVERSE DIRECTION**

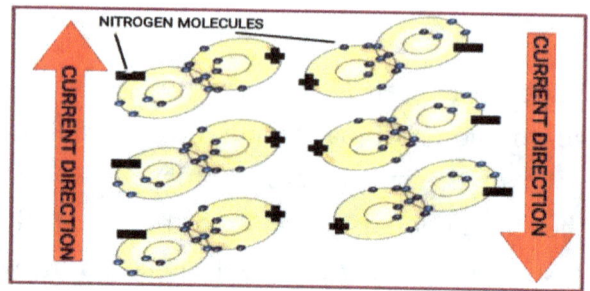

The nitrogen molecules in the air direct their electrons towards the conductive cable, turning these into electric dipoles, leaving the negative pole directed towards the cable and the positive pole on the opposite side, this causes the molecules on both sides to face each other in the center of the cables by the positive pole; producing the repulsion between them.

IF THE CURRENT CIRCULATES IN THE SAME DIRECTION

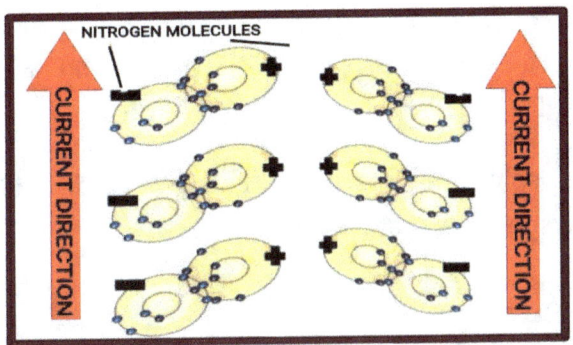

The same situation that occurred with the cables in parallel occurs with currents of counter direction: The nitrogen molecules in the air direct their electrons towards the conducting cable, turning them into electric dipoles, leaving the negative pole directed towards the cable and the positive pole to the opposite side. This produces that in the center of the cables, the molecules of both sides are faced by the positive pole; producing repulsion between them. There should be a theoretical repulsion, but experience tells us that what actually occurs is an attraction. CONCLUSION, BOTH SCHEMES RESULT IN A REPULSION AND THAT IS NOT WHAT HAPPENS WITH AMPERE'S EXPERIMENT.

To solve this paradox, I show below the personal law that I have developed:

GENERAL LAW OF ELECTROMAGNETIC ATTRACTION

ANGEL'S LAW

Coulomb invented the torsion balance in 1777 to measure the attractive or repulsive force that two electrical loads exert on each other and established in 1785 the relationship that links this force with distance: "The magnitude of each of the electrical forces with which two point loads interact at rest is directly proportional to the product of the magnitude of both loads and inversely proportional to the square of the distance that separates them and has the direction of the line that joins them. The force is of repulsion if the loads are of the same sign, and of attraction if they are of the opposite sign."

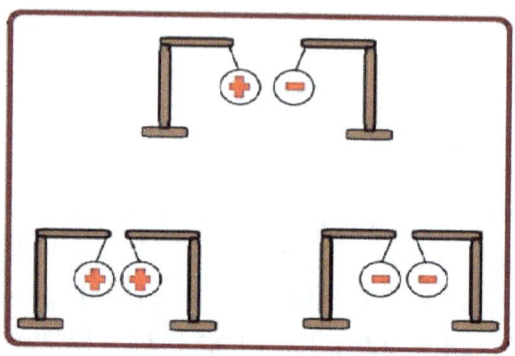

Coulomb's law is valid only in stationary conditions, that is, when there is no movement of the loads. That is why it is called "electrostatic force." Around 1822 Ampere (French) discovers the attraction or repulsion between parallel conductors, which carry current in the same or different direction. Later André Marie Ampère described how two parallel conducting cables through which an electric current circulates in the same direction, attract each other, while if the current of both circulates in the opposite direction, they repel each other.

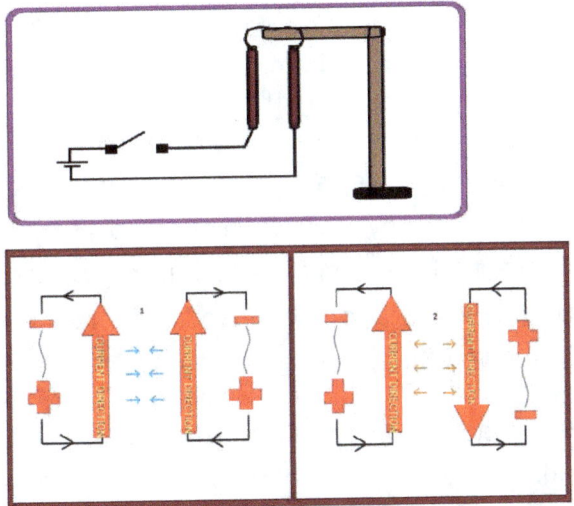

This experiment not only describes the loads in motion, but at the same time it is the base that supports magnetism; moving loads produce magnetism and therefore magnetic attraction and repulsion.

However, it appears that Coulomb's law does not adapt to moving loads and therefore cannot be applied to repulsion or attraction between current cables. This electromagnetic phenomenon is left without a law that could reasonably describe them.

Finding a single law for both experiments, which alone explains the electrostatic forces and the electrodynamic forces, would produce a conceptual electromagnetic unification, since until now, unification between both forces has been solved solely mathematically by Maxwell.

We are going to address the atomic-molecular field. If we consider electromagnetic forces, such as

THE MOMENTUM OF ATTRACTION BETWEEN ATOMS AND MOLECULES TO CAPTURE ELECTRONS, the consequence is that IN ORDER TO PERCEIVE THIS ATTRACTION CONTINUOUSLY, THE EXISTENCE OF A CONTINUOUS FLOW OF ELECTRONS AT THE MOLECULAR LEVEL IS NEEDED.

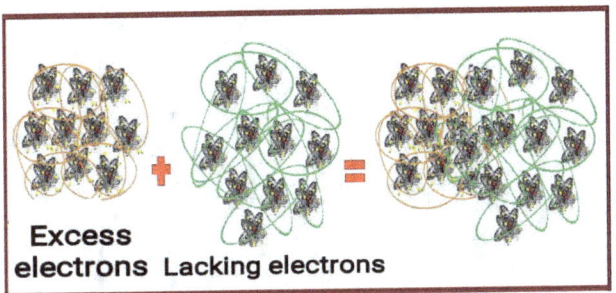

THE SCHEME STUDIED ABOVE WOULD EXPLAIN THIS ATTRACTION.

We can apply this new Law to electric current cables arranged in parallel: 1st TWO CABLES WITH CURRENTS OF THE SAME DIRECTION;

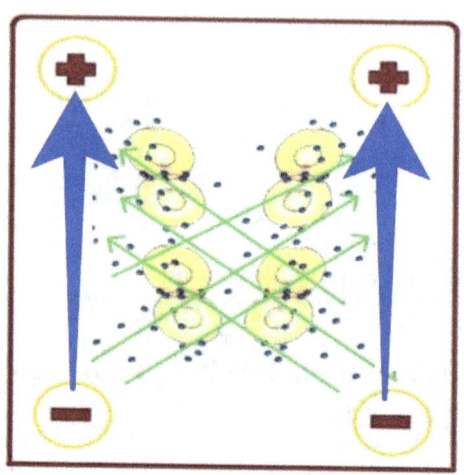

Molecules represented inside the cables are nitrogen molecules, which is the main component of air. Green lines represent the flow of electrons, and as we can see there is not only a flow from the negative to the positive pole of each cable, but there is also a flow from the negative pole of a cable to the positive pole of the opposite, thereby producing the attraction.

Cables also attract electrons from the air molecules on the outside, but from this side the electronic flow is very weak since it does not have a source of electrons as it does on the inside.

2nd TWO CABLES WITH CURRENTS THAT CIRCULATE IN THE OPPOSITE DIRECTION;

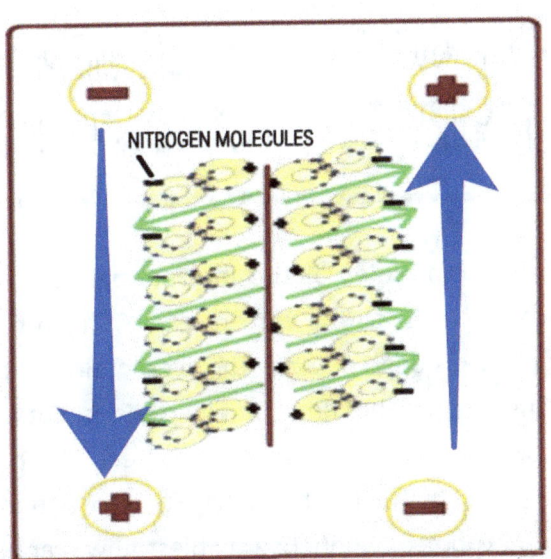

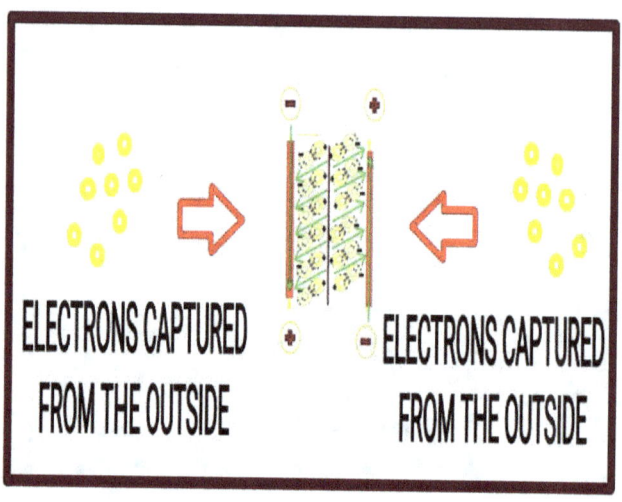

The air molecules of the internal part tend to give electrons to the current cables, from both opposite sides, due to this competition, the transfer and the flow of electrons is cut off, deactivating the attractive effect that is produced by the internal part. The consequence is that both conductive cables now capture electrons from the air molecules on the outside continuously, producing the attraction towards those outside air molecules. This attraction towards the outside was called repulsion under the prism of Coulomb's law, now it is described as attraction towards the opposite side. We can ask ourselves if the current cables really absorb electrons from the surrounding air, then in the case of a single cable without its parallel, it should move when the electric current circulates inside it.

We must bear in mind that in this situation the cable captures electrons from the air in all directions in a uniform way, in such a way that the attraction produced by it is also compensated in all its directions.

NOW LET US SEE IF THIS ANGEL'S LAW OF LOADS IN MOTION IS APPLICABLE TO STATIC LOADS (ELECTROSTATIC).

1ˢᵗ TWO LOADS OF OPPOSITE SIGN:

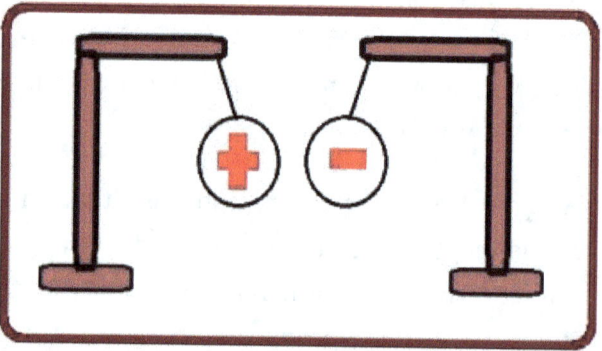

In this case there is a flow of electrons through the air from the negatively loaded sphere to the positively loaded sphere, and the consequence is that both spheres are attracted by the existence of this flow.

2nd TWO POSITIVE LOADS

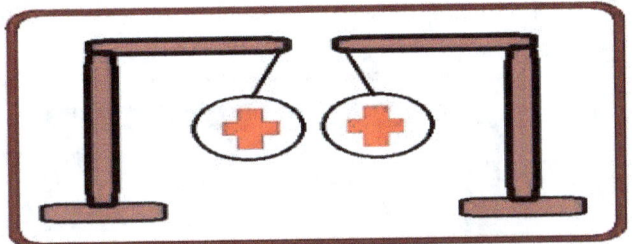

The air molecules on the inside tend to give electrons to the positively loaded spheres located on opposite sides and, lacking a continuous supply of electrons, the emission of the aforementioned electrons towards the spheres ceases and the attractive effect from the inside disappears. However, on the outside, the positively loaded spheres do receive a continuous flow of electrons donated by the air molecules, attracting the balls to the outside.

3rd TWO NEGATIVE LOADS

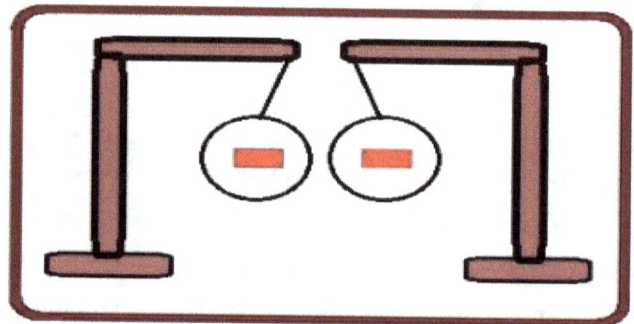

Both loaded spheres emit electrons towards the air molecules on the inside that quickly become saturated with electrons on both sides, the flow is interrupted and the attractive effect disappears as there is no flow on the inside. On the outside, the negatively loaded spheres do emit a continuous flow of electrons towards the air molecules. The balls are attracted to the outside. This law not only unifies attractive electrical phenomena, whether electrostatic or electrodynamic, but it also unifies attractive and repulsive electrical forces into a single attractive one.

With this Angel's Law I describe how in reality the repulsion that we were looking for does not exist, but there is only an electrical force of an attractive nature.

Let me joke; Coulomb's Law is the law of scolded boyfriend and girlfriend; this is if they are boyfriend and girlfriend, they get together, but if they scold each other, the couple breaks up.In other words, Coulomb only found what occurs in nature, which is the approach or repulsion (distance) of loads; but from a deeper and more rational analysis, it is logical to think of a single Attractive Force, not only because of the simplicity of the Theory, but also because assuming Angel's Law, the mystery of 2 Natural Forces is revealed:

ELECTROMAGNETIC FORCE AND

STRONG NUCLEAR FORCE

Indeed; The Strong Nuclear Force is what holds the Protons together in the atomic nucleus, since they should repel each other as they are equal positive loads.If there is no electrical repulsion, we also have no repulsion between protons in the atomic nucleus and the mystery of the Strong Nuclear Force is resolved as non-existent.

Angel's Law is thus endorsed because it correctly explains Ampere's experiment on 2 current cables in parallel, it unifies electrostatics and electrodynamics, it unifies electric attraction and repulsion in a single force of attractive character and justifies the lack of need for the existence of the Strong Nuclear Force.

At this point, we have already addressed 3 of the 4 forces of Nature and only the most difficult and elusive one remains:

GRAVITY

If we drop an object and it falls to the ground, although there is apparently nothing between the object and the ground, we must assume that between the object and the ground there are tiny particles that cannot be observed and that are the cause of producing Gravity.

This is what we call Quantum Gravity and we call the transmitting particle Higgs Boson or Graviton.

There are a number of Quantum Gravity Theories that I am going to briefly critique:

TIES AND LOOP KNOTS these theories in which gravity manifests itself in the quantum world (or that of the very small), through TIES, KNOTS OR LOOPS, I understand that more than attracting what they would do, would be to immobilize bodies in space, and they do not make it very clear how those ties or loops are formed.

TWISTORS: from Roger Penrouse, it seems that the quantum networks that make up bodies are made up of twistors or Tornadoes that are actually beams of light. I don't quite understand how beams of light can produce attraction by gravity, although I do have to admit to this Theory its resemblance in name to the Tornado Theory for Electric Force that I have used in this book.

STRINGS: Strings produce waves, but they do not give an explanation as to how they are the cause of the gravitational attraction mechanism.

After intuiting that none of them addresses the issue of the ultimate reason for gravitational attraction in a rational way:Let us go on to show the personal view on Gravity:

TRAVELING HIGGS BOSONS Previously we had to imagine ourselves in all intergalactic empty space; the existence of a mantle formed by Higgs Bosons; particles so tiny that they are not noticeable even by the best human-made microscopes.

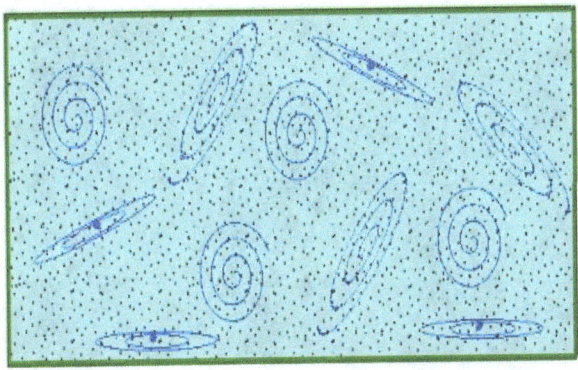

This mantle would reach not only the intergalactic empty space, but also the stars and solar systems and their planets; it would wrap everything around us and the ATOMS that make up matter.

On the other hand, as we have previously described, the protons of the atomic nucleus form Tornadoes, which not only attract electrons to their atomic spiral, but also attract the aforementioned Higgs Bosons. Let us see it graphically:

(protons represented in red in the center of the atomic nucleus). We observe that the atom with its spiral drags the Higgs Bosons inside it. It picks them up on the side where there is higher density and transports them to the opposite side.

These Bosons hit protons of the Atomic Nucleus on the opposite side to which they were picked up, and the hit protons modify their position and spiral in the direction of the side from which the Bosons were initially captured.

This simple and graphic mechanism is what explains Gravitational Attraction.

Higgs Bosons roam the physical medium and are captured by the atom in the area of highest density and hit the Nucleus protons from the opposite side.

The atom is thus capturing on one side and expelling excess Bosons on the other.

Atoms, molecules and materials, thus become binders of Higgs Bosons so that in the vicinity of objects with atoms and molecules, whether they are planets, stars, Black Holes, there is a higher density of Bosons expelled from the atoms and wandering around their surroundings. And that is why I have called this Theory "TRAVELER HIGGS BOSONES THEORY"

These Bosons would also wander through what we call Empty Space, although there, for reasons of distance from atomic foci, they would have a lower density than in the vicinity of interstellar bodies.

Stars, planets and galaxies, move at great speed through the cosmos. (Our own planet circles the center of the Milky Way, traveling enormous distances).

During their journey, these stars and galaxies are agglutinating in their environment Higgs Bosons that they collect with their atomic spirals along the way.

But can this approach marry with

the Experimental Physics that we currently know ?

The first deduced consequence is that Gravity would only affect atoms and not empty space that did not contain atoms. This idea has not been my invention, but it is a consequence of the proposed Theory, and as such we ask ourselves, is it possible that gravity only affects atoms and the world of materials? The answer I give is yes it is possible, although science until now has not contemplated it.

The other fundamental consequence would be that this "Theory of traveling Bosons" would not only explain Gravity but would also reveal one of the main mysteries of current Astrophysics, such as Dark Matter.

DARK MATTER

Dark Matter was proposed by Fritz Zwicky, in 1933

According to Kepler the planets of the solar system travel an elliptical orbit; when approaching the Sun, they accelerate their speed, and when moving away they slow down their speed.

This occurs due to the greater gravitational attraction exerted by the Sun in its proximity and less attraction in the distance.

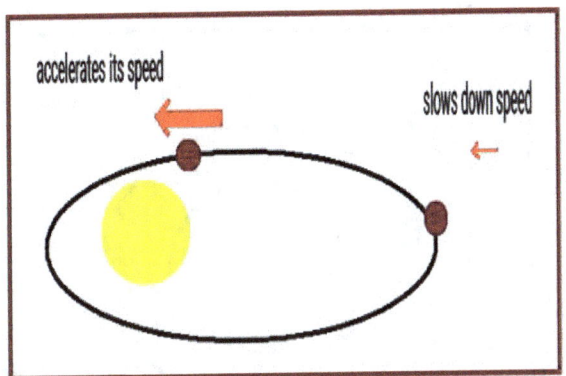

The same should happen in galaxies and stars distant from the center should rotate around the galactic center at a slower speed than nearby ones, since there is a super Black Hole in the galactic center that acts as the sun in the solar system.

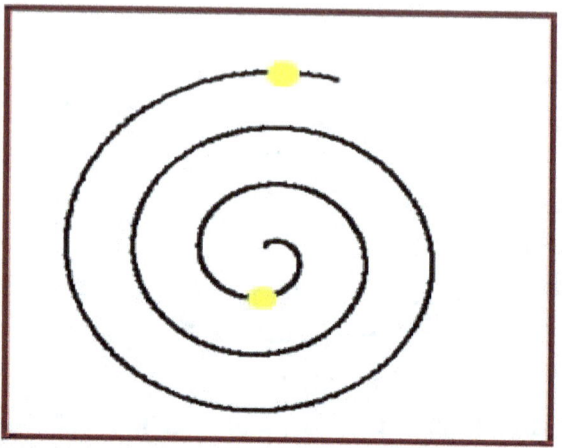

Around the year 1970; The astronomer Vera Rubin was surprised to observe the rotation of the galaxies, stars rotate around the center in reverse in terms of speed: faster those far away and slower those close to the center.

The stars of the galaxy rotate roughly, with the same coordination as a wagon wheel; with all the star radii moving in unison.

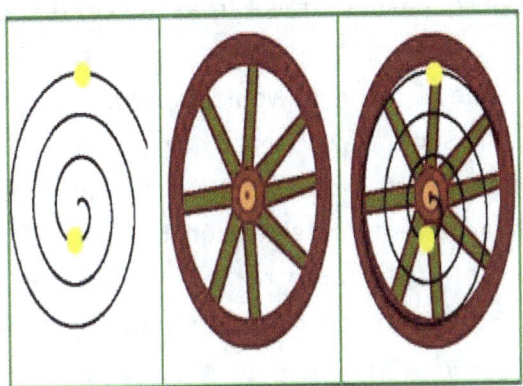

We see in the image that the star close to the axis rotates at a slower speed than the outer star, because it travels more distance than the first in the same time.

The explanation given by the scientific community is that there is a matter in galaxies that are not the stars we see and that produces the necessary attraction for the stars in the galaxy to behave like this (Dark Matter).

"Traveling Higgs Bosons would give a more reasonable explanation for this phenomenon:

Higgs Bosons would travel from the centers of greatest concentration; planets, stars, galaxies, Black Holes and they would blur into empty space. In such a way that they would arrive from one star to another in little concentration, but in an enormous extension of Space. The nearby star would detect this huge expanse of Higgs Bosons in low concentration from the other nearby star and would be oriented towards it.

This would be the explanation of why the outer stars of the Galaxy would move faster, they would simply be oriented and anchored to the Higgs Bosons coming from the closest stars.

This would also occur at the Galactic level: A galaxy would detect the little concentration, but much extension of the Traveling Higgs Bosons coming from the nearest Galaxy and despite the little concentration, it would be oriented towards it or rather, both would be oriented towards the nearby galaxy leaving both anchored to each other in their movement. This would be the ultimate cause of the formation of galactic filaments.

HISTORICAL REVIEW OF GRAVITY

Isaac Newton presented the law that bears its name: "Two bodies attract with a force directly proportional to the product of their masses and inversely proportional to the square of the distances that separate them" and Kepler used Newton's Law to correctly describe the motion of the planets around the Sun.

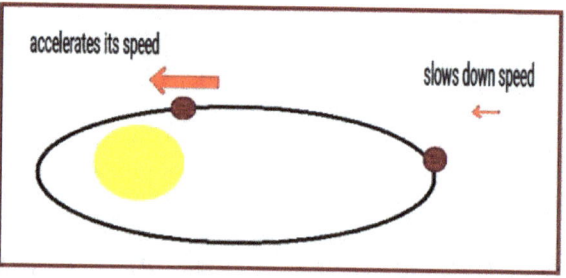

But this system seems to have failed in certain cases, since the orbit of Mercury, the planet closest to the Sun, was more eccentric than that derived from Newton's Law.

This aberration in the orbit was studied by Einstein who with his "Space-Time Fabric" was able to adequately formulate this behavior and thus extend Newton's Law to the proximity of more massive foci.

It seemed that the formulas that described Gravity were already completed with Einstein and could be applied to the entire Universe.

The Gravity described by Einstein was very precise and was constantly being verified with countless experimental observations, but suddenly DARK MATTER appeared on the scene, and since they did not dare to question Einstein's "Space-Time Fabric", they questioned the matter observed in the sky.

It was concluded that there had to be a matter that was not seen that supported the stars in their Galactic motion.

This is obviously not the case because DARK MATTER SIMPLY DOES NOT EXIST.

But, what led the Scientific Community to postulate its existence?

Simply the way to concoct Newton's and Einstein's formulas for gravitational attraction:

Science should have this evolution: first the ultimate causes that produce the Forces are investigated and then formulas are made.

In the case of Newton or Einstein, they did it the other way around, first formulas have been made and then they do not match the reality that nature shows us, since the ultimate causes of such Forces are completely unknown.

However, TRAVELER HIGGS BOSONS THEORY gave me the surprise of solving the problem of Dark Matter, without thinking a priori about such a problem,

The gravitational theory of Traveling Higgs Bosons is endorsed because it explains the mystery of Dark Matter.

The traveling Higgs Bosons would adequately explain the orbit of the planets in the solar system, the eccentricity of the orbit of Mercury and the motion of the outer Stars of the galaxies, as well as the motion of the Galaxies forming galactic filaments.

In other words, they would encompass Newton's Laws, Einstein's General Relativity, and Dark Matter.

Once I have ventured the ultimate causes of gravitational attraction, I can proceed to the creation of a new Law of Universal Gravitation:

"A body is attracted to the surrounding environment with a force directly proportional to the resultant of the density of Higgs Bosons that it suffers around it."

But, following the same reasoning used, "Traveling Bosons" may be censured for not giving an explanation to the "Dark Energy."

So let us move on to the next chapter:

DARK ENERGY

Are the stars of the sky the only matter that exists in the Universe? (not counting Dark Matter).

The stars and galaxies that we observe in the sky, according to astronomers would only account for 4% of the matter existing in the Universe; and a 73 % of non-visible matter would be formed by the so-called DARK ENERGY. BUT, WHAT HAPPENS WITH THIS 73% OF INVISIBLE MATTER.

We are previously going to study the colors:

Why do we see different colors? The reason is the frequency of the light waves that reach our eyes.

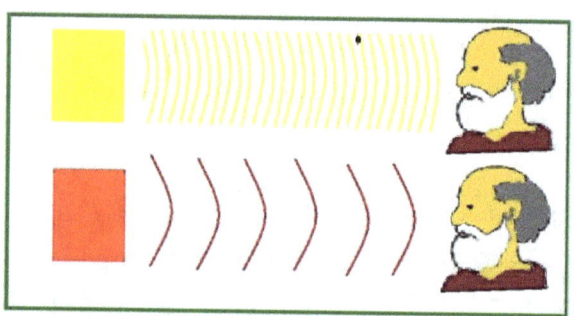

The light waves in the upper yellow box reach us more frequently than the red waves that reach us more widely spaced in time (less frequently).

This frequency in light waves is what differentiates the different colors.

In 1929, Edwin Hubble, the American astronomer who gave the space telescope its name, observed distant stars and galaxies from the Mount Wilson telescope and realized that their light, instead of yellow or white, came in a more reddish hue. In other words, their waves arrived with less frequency than they should carry.

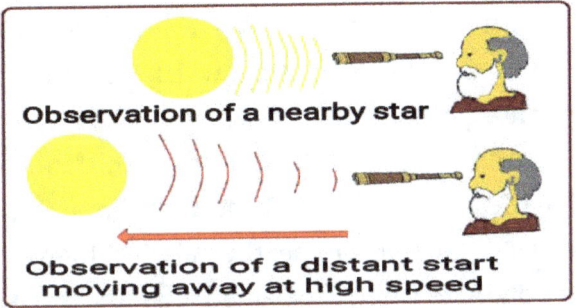

He come to the conclusion that this change in color and frequency occurred because the galaxies were moving away from us at great speed and consequently of the speed of light of the galaxies directed to us, we had to subtract the speed at which the galaxies are moving away in the opposite direction.

In this subtraction was where the waves lost the frequency that they should initially have and that is why they arrived with that reddish hue, it was the so-called Doppler Effect.

The entire Universe was expanding at great speed, it is as if we were inflating a balloon.

If we project the film about the expansion of the Universe in reverse, we will find an initial moment of that expansion, which would be the great explosion or Big Bang.

On the other hand, the more distant the galaxies were; the more its red-shifted waves arrived, however, in nearby galaxies, there was hardly a redshift; it seemed that the galaxies not only are moving away from each other but are accelerating their speed along the way.

But for there to be acceleration, the initial inertia of the Big Bang is not enough; we need something to hit the gas.

That fuel needed to accelerate is what we call Dark Energy.

According to Einstein's Theory of Relativity, energy is equivalent to mass $E=mc^2$, or in other words, mass is the fuel we need for the Universe to expand rapidly.

For accelerated expansion to take place, a mass-energy is necessary; or matter in the Universe much higher than what we observe with the naked eye, we would need a matter-energy that would account for 73% of the total matter in the Universe.

But, where is that 73% of invisible matter-energy? The answer to this question would be what we call the "Enigma of Dark Energy."

However, there is an explanation to the Enigma of Dark Energy: and it would come from the hand of what in physics we call "Standard Model": which affirms the existence of a tiny particle that carries gravity and that today we call the Higgs Boson. The entire Universe would be full of these Higgs Bosons, since the force of gravity is manifested throughout them.

In the vicinity of Black Holes, there would be a large number of Higgs Bosons, and in empty space, there would be a minimal number of these Higgs Bosons. The Higgs Boson would be present as a blurred blanket that would cover the entire Universe.

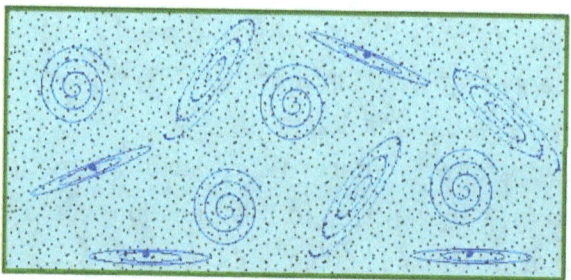

In the image we will represent this blanket that covers the Universe in celestial blue, and the black points would be the Higgs Bosons. In the empty space there would therefore be an invisible fog with a low concentration of Higgs bosons.

We must consider light as a wave (Young's experiment) that travels through a medium; in this case we can consider the medium as the Higgs Boson fabric blurred in our known Universe.

LOSS OF FREQUENCY OF PROPAGATION OF LIGHT WAVES

The disturbance caused by a wave is propagated by collisions of the particles that make up the medium, in this case Higgs Bosons when transmitting their thrust each collides with the neighboring one, in each of these collisions some kinetic energy is lost, and this gradual loss of kinetic energy of the Bosons, ultimately translates into a loss of frequency in the propagation of light waves.

In the image we see how the outer water waves that have traveled further are widening and lose frequency with respect to the inner waves.

Then, light waves would suffer a very slight loss of frequency when propagating through Higgs Bosons medium, but this slight loss of frequency would accumulate over the space of millions of light years, which would gradually slow down the frequency, causing us to observe them red-shifted in their spectrum upon reaching Earth.

close star observation with Higgs Bosons fabric on blue background.

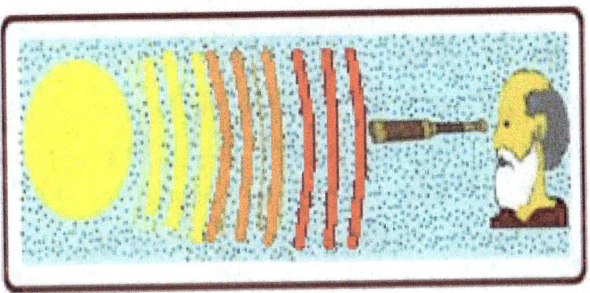

distant star observation with Higgs Bosons fabric on blue background.

Decrease in frequency of light waves would cause them to redshift. An apparent Doppler Effect would occur.

This gradual "frequency slowdown" of the waves would explain and constitute the proof why the most distant galaxies apparently move away with greater speed than the closest ones; light waves lose frequency for a longer way until they reach the Earth.

The consequences of this interpretation would suddenly explain 73% of the matter-energy that we lack in the Universe: this matter would not be necessary because the Universe would not expand rapidly.

However, it would leave the Big Bang orphaned of its most important test, since the expansion of the Universe would be in doubt.

As there is no accelerated expansion of the Universe, we no longer need the engine of this expansion: Dark Energy. The existence of a static or still Universe would not be proven either, it would simply be necessary to take new measurements taking into account this effect - frequency braking of waves.

Ockham's razor tells us that between two possible solutions the simplest is the correct one.

Hubble's prescribed solution to redshift leads us to Dark Energy and to the unknown of where 73% of the matter is in the universe and to the possibility of the existence of repulsive gravity.

The "Braking Waves" solution does not require the existence of Dark Energy, nor 73% of matter

disappeared in the universe, nor a repulsive gravity.
Choose for yourself the solution to this enigma.

INTERPRETATION OF THE PERIODIC TABLE

Let us see how the Electro-atomic Theory of Tornados could interpret the Periodic Table of the elements.To do this, we added a new concept that is "Crown" to the Tornado.

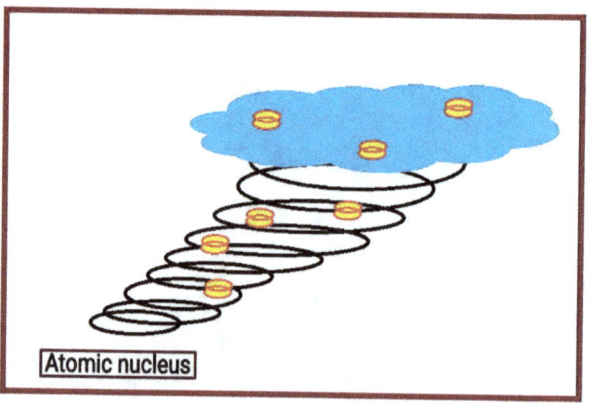

The crown in the blue drawing would be the outermost part of the Tornado where electrons with less attachment to the atom, and therefore, with greater possibility of combination would be.

Combination would occur when another atom approaches, the combinable electrons of the Crown would bypass both the atom itself and the atom that is approaching.　　As we have previously explained in molecule formation (the tornado here is represented with an atomic spiral).

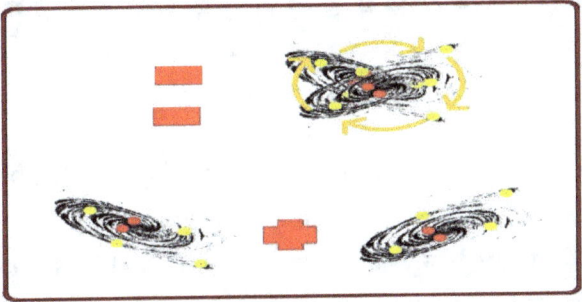

Next we show a scheme on the evolution of the Crown in the atom:

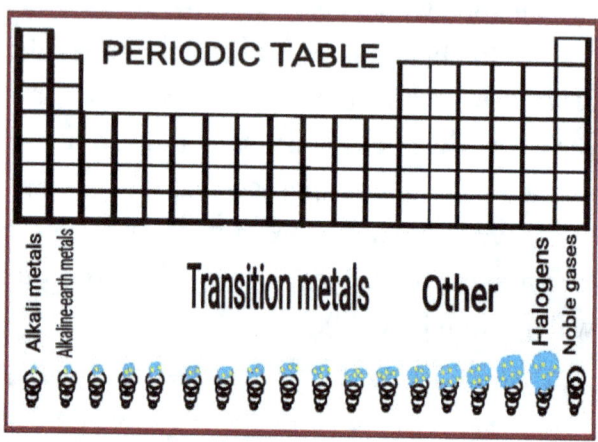

We observe that in the right column (Noble Gases) they do not have a Crown (in blue) and therefore they lack combinable electrons, and do not bond with other elements.

If we add a proton to the Noble Gas tornado; the tornado would gain a little enough strength to start creating a fledgling crown (Alkaline column).

If we add one more proton to the Alkaline, the tornado gains a little more strength and can keep a combination crown a little stronger.

This process continues with the Transition Metals, walking to the right of the Periodic Table, until reaching the Halogen column that maintains the strongest combination crown of all the groups.

If we add one more proton to the Halogens, the atomic tornado will become a little stronger, enough in this case to absorb the combination crown in the tornado, the atom has been converted back into a new Noble Gas without a combination crown.

From here the cycle repeats itself.

The elements on the left side of the table (Alkaline) maintain a weak crown with a low number of combinable electrons, and therefore do not have enough strength to combine by themselves with other elements of the Table.

The elements of the middle left part gradually acquire more combinable electrons in the Crown by adding protons to the nucleus that give more strength to the spiral and the Crown, and therefore maintain a greater capacity to combine both with elements of the right part, as with elements on the left side of the Table.

The elements of the right part through the addition of any additional proton, give more strength to their tornado and crown that acquires more combinable electrons in it than the elements of the middle part of the Table, forming a strong crown, capable of bonding with elements from any area of the Table.

But its strong crown is not enough to combine with the Noble Gases that do not provide a combination crown, and therefore have great difficulty to bind with more elements.

INTERPRETATION OF SPECTRAL LINES

The German scientist Fraunhofer invented the spectroscope, with which we could observe the spectrum of elements.

The light when passing through an optical prism decomposes into colors forming a spectrum.

If we make white light pass through a substance before passing through the prism, only those wavelengths that have not been absorbed by said substance will pass and we will obtain the absorption spectrum of said substance. If we filter light through a gas such as hydrogen as we see in the image:

We obtain a spectrum with black lines that we call absorption spectrum because hydrogen absorbs wavelengths located on the dark lines of the spectrum. Now if we heat a gas to a high temperature, for example hydrogen.

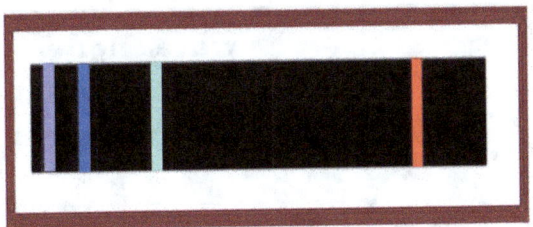

A dark background spectrum is produced with colored lines right in the same place where hydrogen was previously used as filter and produced dark lines.

We call this spectrum emission. And the lines correspond to hydrogen emissions only at certain wavelengths. Each element in the gas state absorbs the same wavelengths that it is capable of emitting. How can we explain these facts using the Atomic Tornado model?

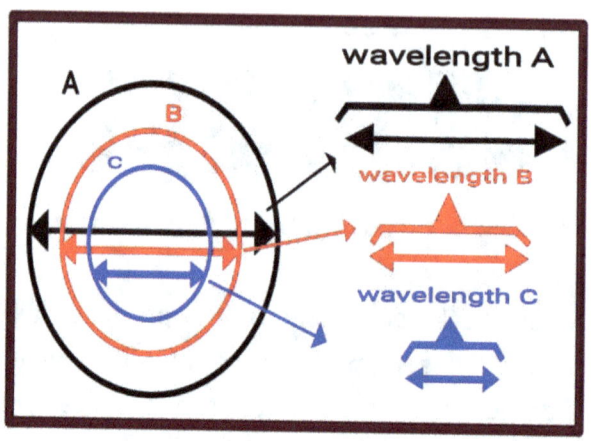

In the image, the atom is a circle that represents the atomic spiral formed by the tornado. What the spectral lines show is the product of the circumvallation of negative particles around the atomic spiral, and why do I say negative particles instead of electrons, because the spectral lines. For example, in the case of Hydrogen, exceed the number of electrons that was given to hydrogen by convention; "1" we must understand that 1 electron is actually the unit of the electric load of the atom and that it neutralizes the load of 1 proton, but we can consider that this unit of load can be obtained not with just one particle, but with several.

Hence, in this study we are going to use the term "Negative Particle" instead of electron.

In the image we show the orbits in which negative particles can move; A, B, and C.

The negative particle "A" makes a bypass along its orbit, producing a wave. As the orbit is wider, the wavelength is also greater and assuming similar speeds within the atom for negative particles, the frequency would therefore be lower for particle A.

For particle B, the orbit would be smaller and as the circumvallation is smaller, the wavelength of generated would be shorter and the frequency higher.

For particle C, the orbit and wavelength would be smaller and the frequency even higher.

Negative particles describe an orbit that, depending on their size, cause a certain wavelength. When this particle receives light, it neutralizes precisely the same wavelength that it generated with its sway inside the atomic spiral, causing an absorption line in the spectrum.

Now if we heat the gas, we cause the negative particles to move or vibrate.

This vibration is transmitted to the medium, causing an emission line at the same wavelength than the particle generates with its orbit.

John William Nicholson, in 1912, suggested that electrons perhaps orbited around a supposedly positive nucleus in orbits whose angular momentum was a multiple of Plank's constant.

He imagined the atom as a small solar system, the radiation of the spectrum was due to the vibration of the electrons within its orbit, undoubtedly Nicholson opened the way of interpreting the spectral lines, taking into account the different orbits of the electrons in the atom, although he did not specify how the absorption lines of the spectrum were formed. Next we are going to make a comparison with one of the few photos that we have of atoms with the spectral lines generated.
"Ytterbium photo of atom obtained at Griffiths University in Brisbane, Australia."

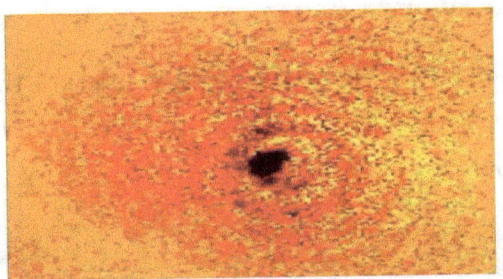

We observe the negatively loaded particles spin forming spherical orbits around the nucleus. If we compare the photo with its spectrum.

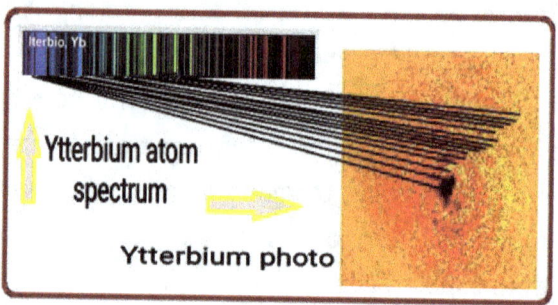

We observe a simple and elegant explanation of how the atom works.

It would be the negative particles around the nucleus that form the lines of the spectrum of each element.

LIGHT

According to Newton, light is composed of tiny particles emitted by luminous bodies.

Christian Huygens, described light as waves that propagate through a universal medium that fills space.

Light as a wave was confirmed by Thomas Young's Double Slit experiment.

Light as a particle was confirmed by: Plank's "Black Body", by Einstein's photoelectric effect and by "Compton Effect."

Plank when studying the effects of the "Black Body" uses a constant (quantum-package), and he himself admits that this effect is due to the interaction of light with atoms and molecules.

Personally, I support Plank's comment since in the 3 fundamental supports of light as a particle, (Black Body - Photoelectric Effect - Compton Effect), an interaction of light with atoms and molecules occurs previously, and as we know, atoms have electrons, protons, leptons, quarks, etc., which are particles and attribute particle qualities to the Light itself.

We have also described in this book how empty space is not empty, but there is a medium formed by Higgs Bosons (travelers), which could provide light with the medium necessary for its propagation in space as a wave.

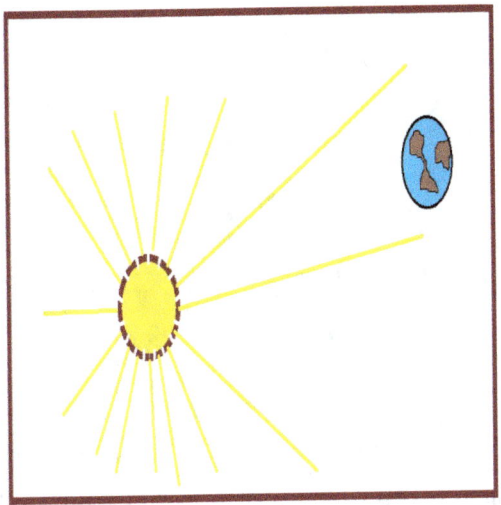

If starlight were particles, rays would open in space and their light would not reach the Earth.

ENERGY

Particles with their movement or vibration transmit this situation to the environment.

This is what we would call energy.

For Einstein, $E=mc^2$ or in other words, there is potentially a huge amount of energy inside the mass.

This formula (not yet proven) visualized through the prism of the Theory of "Tornadoes" and "Traveling Higgs Bosons", perhaps should have another structure:

E = ac² where "a" means "atom"

Interpretation: Inside the atom there is potentially an enormous amount of energy.

Intellectual Property Registry, Madrid,Spain

files

12/rtpi-008541/2008 (10-20-2008)

M 008173/2011 (10.25.2011

M000960 / 2013 of 5-2-2013)

e mail: angelperez94@gmail.com

Copyright   "Espiral"   123RF   invoice
8TK371248B118481D

www.ingramcontent.com/pod-product-compliance
Lightning Source LLC
Chambersburg PA
CBHW070302220526
45465CB00004B/1708